C0-BIV-604

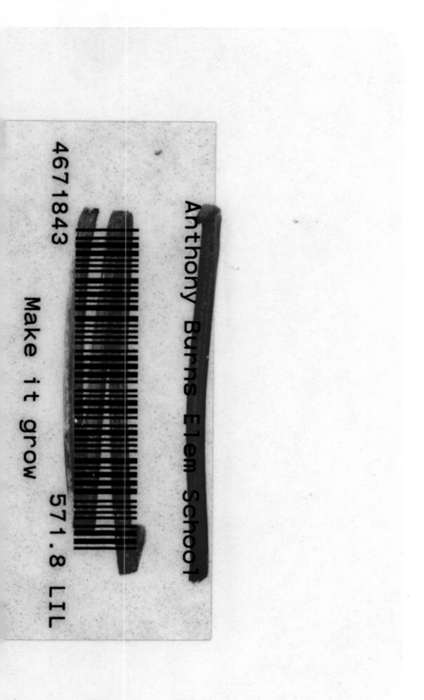
Anthony Burns Elem School
4671843
571.8 LIL
Make it grow

Read and Do Science

Make It Grow

Written by Melinda Lilly
Photos by Scott M. Thompson
Design by Elizabeth Bender

Educational Consultants
Kimberly Weiner, Ed.D
Betty Carter, Ed.D
Maria Czech, Ph.D
California State University Northridge

Rourke
Publishing LLC
Vero Beach, Florida 32963

Before You Read This Book

1. Compare what you know about plants and animals. How are they different? In what ways are they alike?
2. What type of foods do people eat? What might plants use as food?

The experiments in this book should be undertaken with adult supervision.

For Kirk

—S. T.

Library of Congress Cataloging-in-Publication Data

ISBN 1-58952-637-6

Printed in the USA

Table of Contents

They're green, they're growing, and they're all around us!

How did this get started?
Seeds and other things like them: **bulbs,** even **spores.**

This could get dirty. They'd like that. Dirt, sunlight, air, comfortable temperatures, and water are what they need to live.

What are they, aliens?
No, they're plants!

Let's grow plants to look like aliens!

Alien Plants

What You Need:

- Half of an eggshell and an egg carton
- Markers
- Two cotton balls
- Rye grass seeds (rye seeds like warm temperatures*; if it's too hot or cold, choose another type of grass seed instead)

- Potting soil

- Spray bottle with water

- Safety scissors

*From 60 to 85 degrees Fahrenheit (16 to 29 degrees Celsius)

Put two wet cotton balls inside the eggshell.

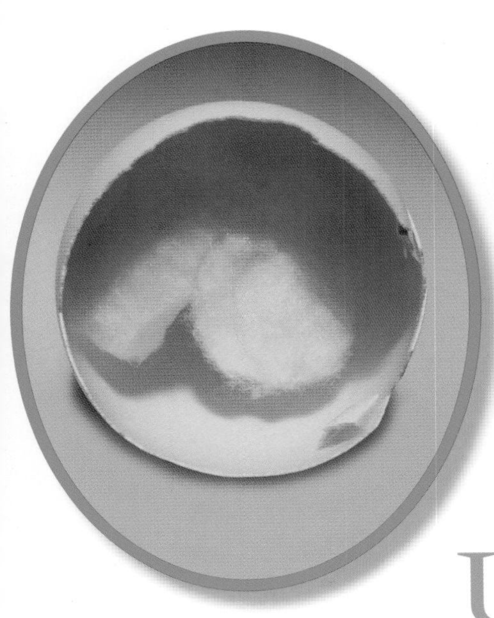

Use markers to gently draw an alien face on the eggshell.

Put a handful of grass seeds on the cotton.

Put the spray nozzle inside the egg. Spray with water.

Sprinkle a handful of **potting soil** on the seeds.

Spray water inside the egg.

Ask an adult to help you to cut out an egg holder from the carton.

Place the eggshell in the egg holder.

Set your alien eggshell in a protected, sunny spot, such as a windowsill.

Spray water inside the egg each day.
Look for grass after a week.

Why Did the Seeds Germinate?

Here are some clues in this **limerick:**

> There once was an alien with seeds.
> Dirt, water, and light were its needs.
> It started out bare,
> Then grew grass as hair,
> Now to keep itself neat, it weeds!

Yummy Plant Food

Plants make their food by a process called **photosynthesis.**

I'd rather eat an orange, thanks.

What's the recipe for photosynthesis?
Plants use sunlight, water, a gas called **carbon dioxide,** and **minerals.**

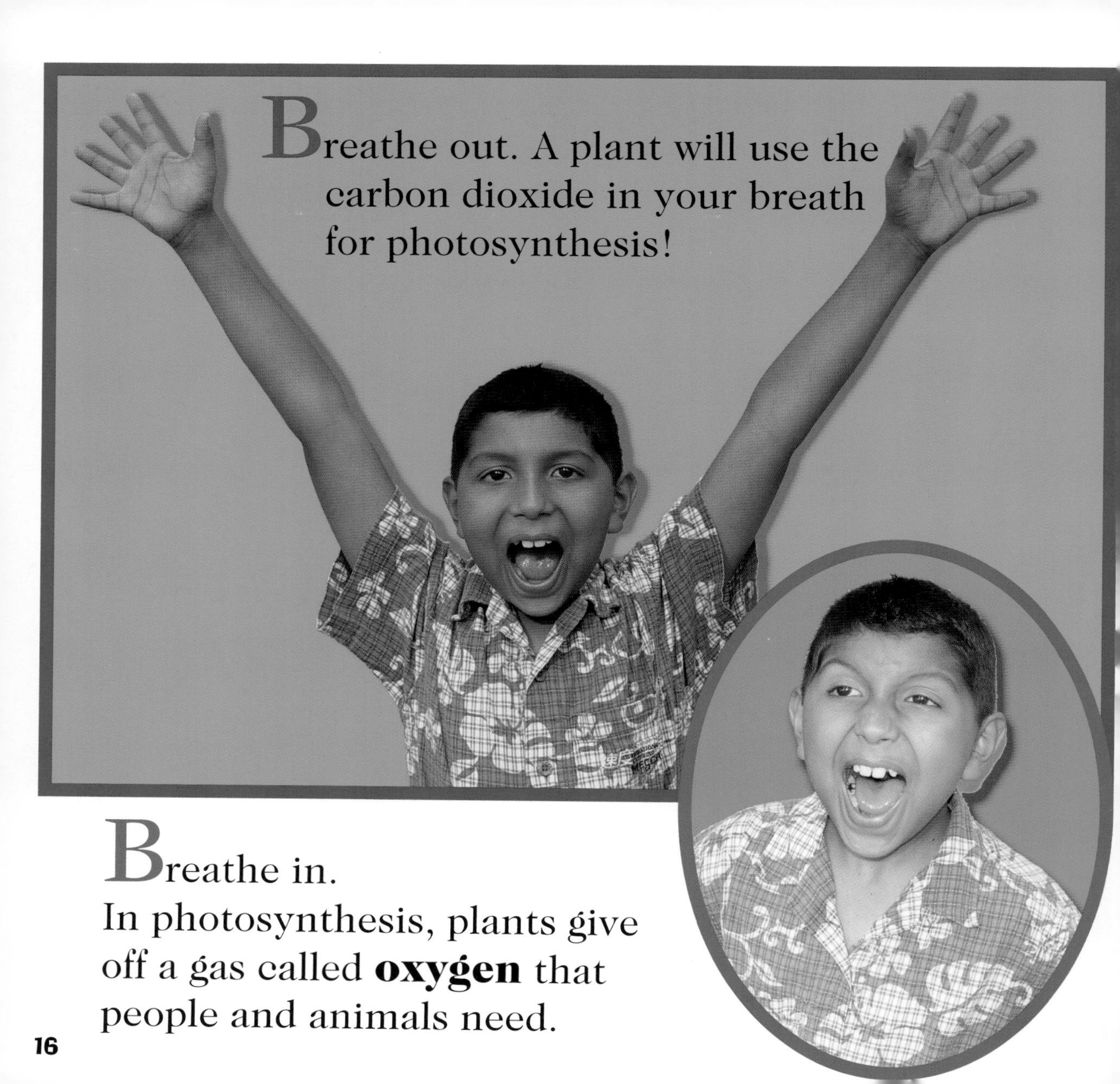

Breathe out. A plant will use the carbon dioxide in your breath for photosynthesis!

Breathe in.
In photosynthesis, plants give off a gas called **oxygen** that people and animals need.

How do plants get minerals? Their roots suck minerals out of the soil.

See lettuce roots!

Lettuce Roots

What You Need:

- Clear baggie
- Plastic bag that is larger than the baggie
- Potting soil
- Safety scissors
- Lettuce seeds (lettuce seeds grow best in cool weather*; if it's too hot, substitute Swiss chard or basil)
- Spray bottle with water
- Newspaper

*From 50 to 70 degrees Fahrenheit (10 to 21 degrees Celsius)

Snip a tiny hole at both bottom corners of the baggie.

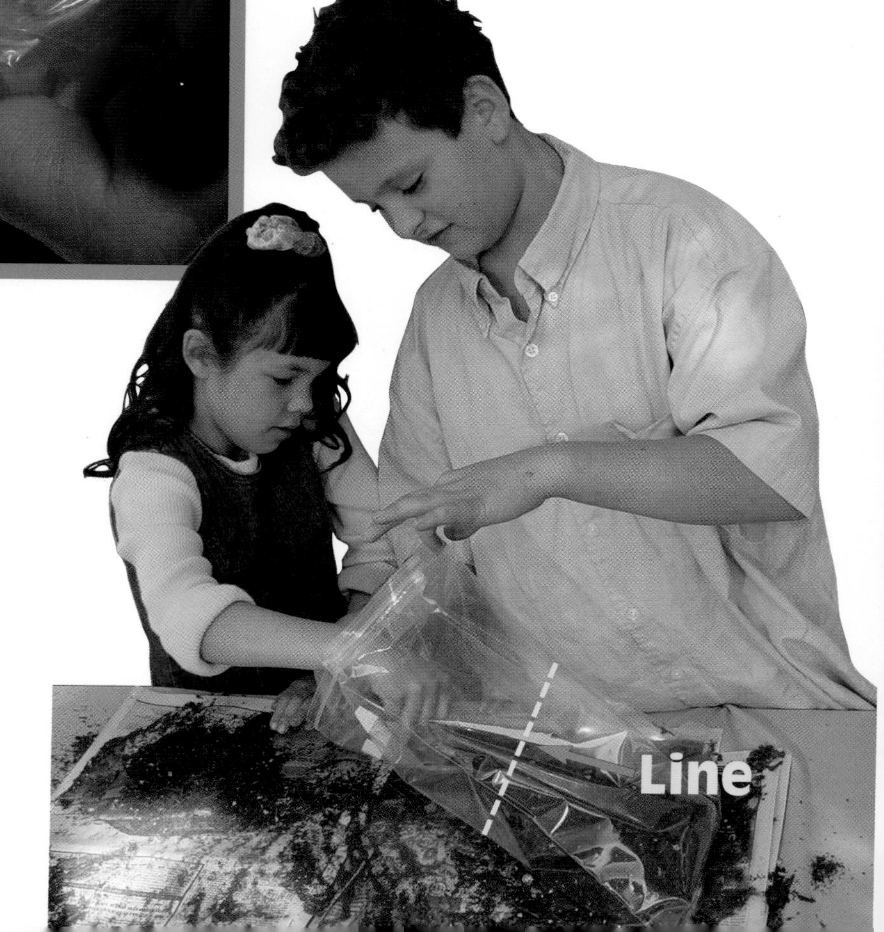

Set the bag on newspaper.

Fill the baggie up to the line with potting soil.

Spray water on the soil until it is damp.

Sprinkle twelve seeds on the soil.

Sprinkle a handful of soil on the seeds.

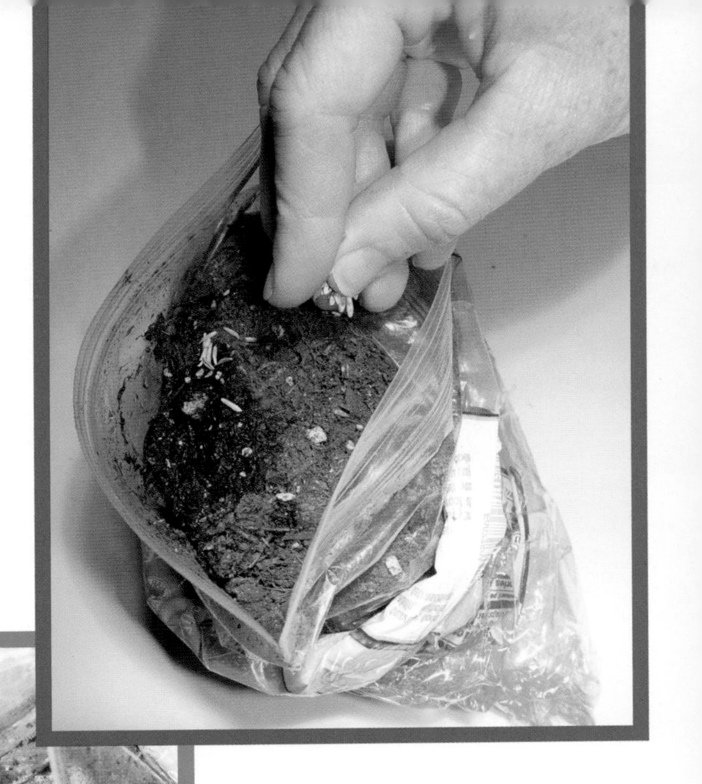

Spray water on the soil.

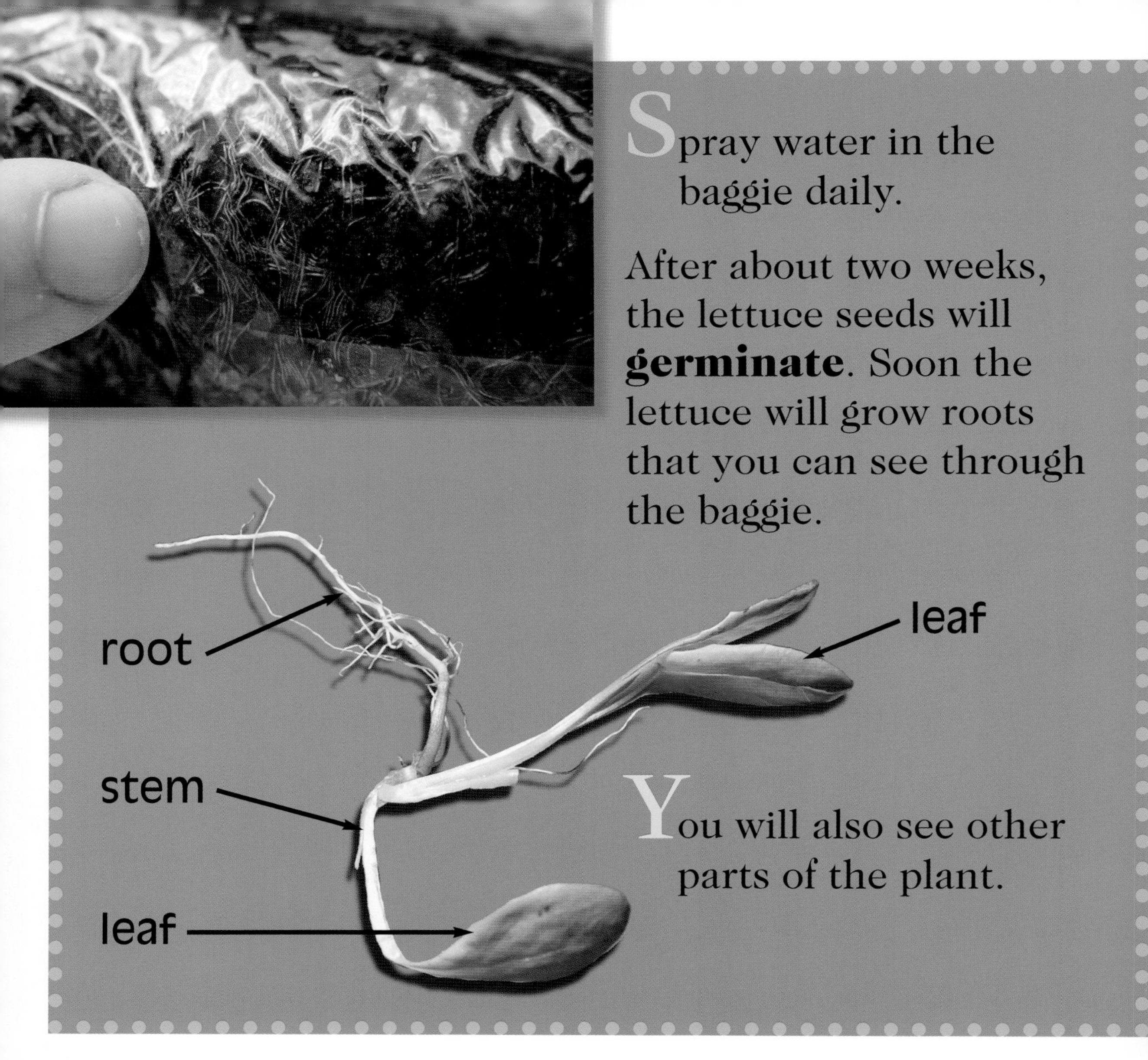

Spray water in the baggie daily.

After about two weeks, the lettuce seeds will **germinate**. Soon the lettuce will grow roots that you can see through the baggie.

You will also see other parts of the plant.

More and More Plants!

When the lettuce gets old, it will flower. After it flowers, it will make seeds. If you plant those seeds, what will happen?

With moist dirt, air, and a cool place to grow, the seeds will germinate. Then more and more plants will grow!

Glossary

bulbs (BULBZ) — buds from which plants grow, they are planted in the ground like seeds, and leaves come out of their tops and roots from their bases

carbon dioxide (KAR bun die OK sied) — A gas that is breathed out by animals and used in photosynthesis by green plants

germinate (JUR muh nate) — the process by which a seed, bulb, or spore puts forth shoots, to sprout

limerick (LIM ur ik) — a silly rhyme in which the first, second, and fifth lines rhyme with each other and the third and fourth lines rhyme with each other

minerals (MIN ur ulz) — Natural substances that are essential to the health of plants, animals, and humans

oxygen (OK see jun) — A gas that that is necessary in breathing and is a byproduct of photosynthesis

photosynthesis (foe toe SIN thu sis) — the way plants make food using sunlight, water, and carbon dioxide

potting soil (POT ting SOIL) — a good dirt mix for growing plants

spores (SPORZ) —seeds or germs

Take It Further: Growing Salad

When the lettuce sprouts two leaves, choose plants from the baggie that you want to continue growing. If the lettuce grows in a sandwich bag, choose one plant. If your bag is gallon sized, choose three.

Gently remove the other plants and some of their surrounding soil. Plant them in a protected spot outside or put them into pots and add more soil.

When the leaves are large enough for a salad, pick a few. Wash them, add dressing, and enjoy!

Think About It!

1. What happens first? Look at these words and think about what would be the right sequence in the life of plants.

   Make flowers

   Sprout leaves and roots

   Germinate

   Make seeds

2. Can you predict what would happen if you did not provide the seeds with moist dirt and a comfortable place to grow?

3. How are grass and lettuce alike? Do people use them for different purposes? Compare and contrast grass with lettuce.

Index